Where Water Meets Land

by
LUCIAN NIEMEYER

Published by
YORK GRAPHIC SERVICES, INC.
1996

*Dedicated to
my wife Joan who shared
this incredible journey with me.*

Copyright © 1996 by Lucian Niemeyer

All rights reserved.

Printed in the United States of America by
York Graphic Services, Inc.,
3600 West Market Street, York, Pennsylvania 17404

Distributed by LNS Art, 658 Mount Road, Aston, Pennsylvania 19014

ISBN 0-9651966-0-7

Cover illustration: Whooping cranes in flight,
Grus americanus, endangered. Aransas National Wildlife Refuge, Texas.

Photography by Lucian Niemeyer

Where Water Meets Land

Preface

I was born in Germany in 1937, and my parents fled to the United States just before the War. My father was a professor and we settled at Princeton University. We moved to Oglethorpe University in Atlanta, Georgia, in 1945. There in school, I first learned about the great mysterious Okefenokee Swamp. The alligators, swampers, Ivory-billed woodpeckers, and the fabulous Whooping cranes (which I later learned were not there) caught my imagination. In Chamblee Elementary School, my fourth grade teacher Mrs. Burson, gave me my first bird book. The impressions of these two events were indelible and were to result in three studies in that I have spent many years completing. *Long-Legged Wading Birds of the North American Wetlands,* Stackpole Books 1993, with a text by Dr. Mark Riegner, is a simple story of wetland habitat which uses a complete study of the long-legged waders as actors in a unique environment. *Okefenokee: Land of the Trembling Earth* is a most complex wetland and its story is written well by Dr. George Folkerts of Auburn University. This volume, *Where Water Meets Land,* York Graphic Services 1996, breaks loose of the discipline of these intense studies and through pictures allows the reader to imagine broader areas of wetland discoveries that were not possible to depict in earlier studies.

I was schooled in more traditional subjects at Indiana University and the University of Notre Dame. Not being able to find my niche, I later studied photography for which I received a degree at the Famous Photographers School in Connecticut in 1972, resulting in a career that I could not anticipate while serving in the automobile industry with Volkswagen of America, ultimately as National Sales Manager and later as owner of a Volkswagen dealership. Meanwhile I continued to enjoy and work at my photography skills. In 1987, my wife Joan suggested that I end my business career and work on photographic studies. From my early work came *Chesapeake Country,* Abbeville Press 1990, with a text by Eugene Meyer; *Old Order Amish,* Johns Hopkins University Press 1993, with a text by Dr. Donald Kraybill; and *Shenandoah: Daughter of the Stars,* LSU Press 1994, with a text by Julia Davis. Permanent collections of the photographs featured in *Chesapeake Country* and *Old Order Amish* are located at the Mariner's Museum in Virginia, and at Elizabethtown College and Lehigh University in Pennsylvania.

My studies commence with a passion that leads to intensive research on the topic. Then I lay out a plan that encompasses two years of field work. Halfway through, I engage a text writer that has a similar feeling about the topic and is an expert in the field. We then start to mesh our messages culminating in a whole study. I use Leica cameras without filters, artificial light, or darkroom manipulations. I do use long lens. My films are usually chosen to represent the image as I saw it, without embellishment or enhancement. I feel that I am recording as accurately as one can, observing nature in the field. The one artistic license that I use is my positive feeling for beauty, so I leave out the trash, power lines, human signs, and other detritus which tend to confuse my sense of a powerful image. I hope you enjoy the story.

My work would not be possible without the encouragement and support of my wife, Joan, to whom all of the wetland studies are dedicated. I am also grateful to my friend Leo Lamer who encourages me and to Ray Chronister, Laura Skinger, and the staff at York Graphic Services who solicited the book and gave it life so carefully. All of this would not be possible without my maker and the wonderful Earth that He placed in front of my cameras. I am so thankful.

LUCIAN NIEMEYER
ASTON, PENNSYLVANIA, 1996

Introduction

"*Where water meets land*" is a joining of two massive environments on this earth. It is an edge on the planet where nature crosses, allowing the land residents to feed and bathe in the water and water creatures to sun themselves and incubate safely. It is a unique place where specially adapted creatures join in the plethora of life created by an intense amount of sun, food, and water—the sustenance of all life on earth. It is a place where humans can study the food chain easily. It is a place where insects, humidity, and the active, bountiful animal and plant life can keep a human busy in self defense, yet in awe of the sheer beauty of the natural order at close range. A place where humans may be able to recognize the niche that they occupy while they spend their short eighty years or so on earth. It is a remarkable place with names that ring out mysteriously and resonant: *Atchafalaya, Okefenokee, Everglades, Assateague, Anahuac, Mississippi, Loxahatchee,* and so many more, often names that were suggested by old Americans who inhabited the United States long before the European settlement.

Yet beyond the names, few people in the United States know of the importance and loveliness of these incubators of life: The salt water marshes that range from Maine along all coasts south and then north and west of Texas, whose grasses absorb the punishing energy arriving in storms along the coasts; the rivers that drain east, west, and south, pouring nutrients into oceans and the gulf through large deltas; the river floodplains that absorb the flow of a cycling ebb and flow of showers; the swamps that soak up the rain overflow, which then filters the water as it provides nourishment for dry lands, releasing it slowly like a sponge; the lakes and prairie pot holes that are nature's holding tanks, spreading their moisture slowly and evenly, revealing age-old patterns of distribution and successful adaptation by living creatures; the springs, underground rivers, and subterranean aqua table structure that provide us with life-supporting filtered pure water. It is a marvelously interconnected support structure for life on this earth. We should understand it as well as we know our own home and all of its functioning systems, as indeed the earth is our abode.

In this volume I have not been constrained by the formality of a specific study. Rather, in my images I have let the imagination run, giving you a broad cross section from coast-to-coast wetland observations via carefully selected, purposeful, representative images. Here I ask the observer to take a photographic trip across our magnificent country, visiting rarely seen places, plants and animals of this important edge of "where water meets land." I let a natural portrait explain the enormous beauty of this important environment, which we humans have treated so poorly. We have a history of dismissing them as inhospitable to man and economically inferior for commerce.

It has been a sad chapter, in that this environment of the most productive of life givers, the wetlands, has been reduced in this country by over 50 percent in the last 200 years, and this assault has not stopped as of today. Somewhere we will cross that line where it will directly and forcefully impact us as the stewards of our planet. Hopefully, then, the course can still be healed by the natural order, which quite often is so forgiving. Meanwhile we have to learn to understand our Planet Earth better. I ask, no I beg, for the leadership in our country and the world leaders to understand the balances of a natural order. I hope they would then place into policy a plan, which could be a blueprint for human living, without excesses, compatible with stewardship of an earth that we have inherited from age-old generations. We must allow our children's children to inherit the same home that we have been given. We owe that much.

Pfeiffer Beach, Big Sur, California.

On the overleaf:
A map of the United States showing the locations of the National Wildlife Refuges, lakes, rivers, and bays in which the photographs for this text were taken.

Flamingo in the Everglades National Park at dawn is at the end of the fifty-mile wide shallow Shark River, which drains the area from Lake Okeechobee to the Florida Bay.

Passage Creek — a small stream in Fort Valley, Virginia — drains a mountain area, runs into the North Fork of the Shenandoah River, then into the Potomac, which flows into the Chesapeake Bay, which surges into the Atlantic Ocean.

A Bullfrog, *Rana catesbeiana*, in the Bombay Hook National Wildlife Refuge, Delaware. Tidewater marshes and pools along the coasts provide a buffer between the great water areas and land.

An autumn scene in the middle fork of the Suwannee River, Georgia, which initiates from the Okefenokee Swamp, as does the St. Marys River.

A Sandhill crane, *Grus canadensis,* preens its feathers in Chesser Prairie, Okefenokee Swamp, Georgia. "Okefenokee" interpreted in Native American tongue means "land of the trembling earth."

A Purple gallinule, *Porphyrula martinica*, in Loxahatchee National Wildlife Refuge, Florida, which is part of the large watershed starting at Lake Okeechobee and ending in the Florida Bay.

Bombay Hook National Wildlife Refuge, Delaware.

Spartina grasses in Blackwater National Wildlife Refuge, Maryland.

A female King rail, *Rallus elegans,* beckons to her offspring to find shelter under her wing in Anahuac National Wildlife Refuge, Texas.

A nutria, *Myocastor coypus,* in Atchafalaya Basin, Louisiana. This large rodent was introduced to North America from South America and now populates the whole southern wetland region.

The Merced River in Yosemite National Park, California, runs with snow runoff in the spring, signaling the renewal of life via water and sun.

Grass pink, *Calopogon pulchullus,* on Chesser Island, a natural museum in the Okefenokee Swamp — North America's most complex wetland environment.

River Narrows, Okefenokee Swamp, Georgia.

Black-crowned night heron, *Nycticorax nycticorax,* are found in Chincoteague National Wildlife Refuge as well as throughout the United States.

Whooping crane, *Grus americanus*, in flight over Aransas National Wildlife Refuge, Texas. This endangered population fell to less than 20 and is recovering slowly. There are now approximately 150 wild and 60 captive Whooping cranes, and they are still on the threshold of extinction.

The Atchafalaya River — the mysterious land of the transplanted Arcadians, or "cajuns" — is a floodplain for the mighty Mississippi River. The Army Corps of Engineers, seeking to confine the flooding of the river, have attempted to reroute its flow. Nature, however, is seeking to reestablish itself by altering the course to its original flow. Lake Fausse Pointe, Louisiana.

A Black-necked stilt, *Himantopus mexicanus,* in Merritt Island National Wildlife Refuge, Florida.

River Narrows, Okefenokee Swamp, Georgia. Made up of river flood plains, freshwater springs, prairies (water), batteries, rivers, ponds, lakes, hammocks, bogs, and swamps, the Okefenokee provides a unique study of a wetland environment in the United States.

Grasses in the Loxahatchee National Wildlife Refuge, Florida.

A Least bittern, *Ixobrychus exilis,* in Eco Pond, Everglades National Park. This diminutive heron, a furtive resident of reed marshes, is relatively common, though seldom seen.

A Canada goose, *Branta canadensis*, in Chincoteague National Wildlife Refuge, Virginia. These common migratory geese use one of three major flyways in North America. They winter in lower states while flying north into Canada for the summer.

Canada goose gosling, Chincoteague National Wildlife Refuge.

While the Pacific Ocean coastline, shown here at Big Sur, California, is dramatic with sharp cliffs and distinct edges, the Atlantic and Gulf of Mexico coasts have more saltwater marshes and barrier islands.

A southern bird of the marshes, a White ibis, *Eudocimus albus, imm.*, along the Anhinga Trail in Everglades National Park.

Although normally feeding independently, these Tricolor herons, White ibises, and Roseatte spoonbills convene at a communal feeding on the Merritt Island National Wildlife Refuge each morning and then disburse to more specialized feeding areas.

Storms and grasses provide for a dramatic moment at Anahuac National Wildlife Refuge, Texas. Buffers of the saltwater marshes absorb the shock of the storm, dissipating the energy protecting the mainland. These shallow water wetlands are nature's incubators and protectors for growing wildlife while also providing an effective filter from unwanted debris.

This unique bird — the Limpkin, *Aramus guarauna* — is found only in central and northern Florida. It is shown here in Cross Creek, home of Marjorie Rawlings, author of the classic *The Yearling*. Its habitat is being reduced, and it is in danger of survival as a species.

The American alligator, *Alligator mississippiensis,* is at the top of the food chain in the swamps. Once threatened through overhunting, it has now resumed its large population in the south. It is shown here in Okefenokee Swamp Park, a privately owned park with an educational environmental classroom of the swamp, Waycross, Georgia.

Pond cypress grove near the Sill, Okefenokee Swamp, Georgia. This area was clearcut in the 1900s and dammed in the 1950s. The swamp is now recovering its natural order, providing visitors with an unparalleled look at wetland environment.

A large Pond cypress in Billy's Lake, Okefenokee Swamp.

Florida redbelly turtles, *Chrysemys nelsoni,* with feet outstretched, sun themselves in the late afternoon in the spring in Loxahatchee National Wildlife Refuge.

The Common moorhen, *Gallinula chloropus,* in spring. The mothers with new hatchlings are ever vigilant for predators.

A Pig frog, *Rana grylio,* in Okefenokee Swamp Park.

Billy's Lake, the largest lake in the Okefenokee Swamp. The fresh tea-colored water of the Okefenokee has an unusually high acid content, which limits the mosquito population and gives the swamp its mysterious dark water. Mariners traveled far to obtain the water from the St. Marys River because its high acid content kept it fresh for long periods of time.

Long and narrow, Minnie's Lake provides a slow flow for the Suwannee River as it winds its way through the Okefenokee Swamp.

Yellow-crowned night heron, *Nycticorax violaceus,* in the morning light on "Ding Darling" National Wildlife Refuge, Sanibel Island, Florida.

A Little blue heron, *Egretta caerulea,* in the middle fork of the Suwannee river. In its juvenile stage this bird is pure white; it molts to blue in its second year.

Autumn at Billy's Lake.

A Snowy egret, *Egretta thula,* on Chincoteague National Wildlife Refuge. In flight, this clown of the heron family is distinguished by its yellow slippers.

Little blue heron hatchlings on Pea Island, Delaware River, one of the largest rookeries on the east coast. Rookeries are safe areas in which many species of long-legged wading birds nest together. Usually, they will be found near food sources available in the spring.

The Atlantic Ocean barrier island Chincoteague is home of the feral Chincoteague Ponies, which are rounded up each July by the local fire department. Some of the ponies are sold each year to keep check of the population.

Osprey, *Pandion haliaetus,* in flight. This bird population was seriously diminished by the use of DDT prior to the 1970s, but has since regained its prominence in the food chain.

Osprey landing on a nesting site at Moores Point of View, Chesapeake Bay, Maryland.

A Laughing gull, *Larus atricilla* — an eastern and southern coastal gull — in its breeding garb.

This immature Osprey on its maiden flight from the nest ran out of strength and fell into the Chesapeake Bay. It struggled to shore, where it dried out and flew weakly to its nest to get food from its parents and regain strength.

Black gum, *Nyssa sylvatica,* sometimes called Quichee lime by the local people, is shown in the Sill area of the Okefenokee Swamp. Here an earthen dam was built by the Army Corp of Engineers to maintain a level of water in the swamp. The dam changed the natural order of the swamp plantlife, yet the swamp has adjusted to the new water table.

River Narrows near the Sill with an Alligator holding court.

Golden club or Neverwet, *Orontium aquaticum*. In February, the large yellow groupings provide visual relief from the still grays of winter.

Golden club on the middle fork of the Suwannee River.

Pond cypress in the Chesser Prairie of the Okefenokee Swamp.

Pods of Titi, *Cyrilla racemiflora,* in autumn. This dense shrub is common in the southern swamps.

Pitcher plants, *Sarracenia flava,* on Chesser Island, Okefenokee Swamp.

This Dusky pigmy rattlesnake, *Sistrurus miliarius barbouri,* in newly molted garb, was stepped over by the author prior to being noticed on Billy's Island, Okefenokee Swamp.

A Brown pelican, *Pelecanus occidentalis,* in Everglades National Park.

Low water in Billy's Lake reveals the roots of a Pond cypress in the Okefenokee Swamp.

Deep in the Okefenokee Swamp is Big Water, a lovely lake with Pond cypress making their patterns in the dark water.

The Mute swan, *Cygnus olor,* shown here in Rockland County, New York. These birds were imported from Europe and have grown into a sizable population in the United States.

Grasses on the Chesser Prairie in spring. Water "prairies" in the Okefenokee Swamp are large expanses of high water broken up by islands of more permanent vegetation, some with trees. When the water subsides, these grasslands are crisscrossed by streams leading out of the swamp.

Quite common in the United States, the Great blue heron, *Ardea herodius,* evokes wonderful feelings of the natural order when observed in its habitat — Billy's Lake, Okefenokee Swamp.

White pelicans, *Pelecanus erythrorhynchos,* on Florida Bay.

An American bittern, *Botaurus lentiginosos,* in Anahuac National Wildlife Refuge, Texas.

A female Mallard duck, *Anas platyrhynchos*, in Cape May, New Jersey.

Snow geese, *Chen caerulescens,* are a common sight in the winter on the Chesapeake Bay.

Domestic ducks provide a humorous image as they walk over the ice to assist a frozen-in family member in the Poconos, Pennsylvania.

68

The Muskrat, *Ondatra zibethicus,* is a common rodent of saltwater marshes — Chincoteague National Wildlife Refuge.

Sandhill cranes, shown here in the Chesser Prairie, are large wading birds with a distinctive call that can be heard for miles. Some birds reside in the Okefenokee all year while others migrate to and from Florida.

Seagrove Lake and tower in Okefenokee Swamp.

Lesser yellowlegs, *Tringa flavipes*, feeding in Merritt Island National Wildlife Refuge.

Willet, *Catoptrophorus semipalmatus*, feeding on the shore of Galveston Bay, Texas.

Grasses in the Bay of Florida.

Long-billed curlew, *Numenius americanus,* in Laguna Atascosa National Wildlife Refuge, Texas.

At Cape Hatteras, North Carolina, the Gulf Stream and strong southbound currents meet, creating violent weather conditions that are extremely hazardous for mariners.

The Reddish egret, *Dichromanassa rufescens*, has a completely white form called the White-reddish egret. This bird on Dead Mans Island in Aransas Bay, Texas, displays some of the dimorphism of the species.

Whooping crane in Aransas Bay, Texas.

Sandhill crane in flight near the Platte River, Nebraska, in early spring during migration to the north. This funnel of the flyway creates a special problem because the habitat is changing as water is siphoned off the river to meet agricultural needs.

Water lily with frog.

A Great blue heron in mating plumage in the spring — Blackwater National Wildlife Refuge, Maryland.

This Osprey has been studied from birth in 1994, by the author, as its parents return each year to fledge two chicks on the Chesapeake Bay.

Big Water in the Okefenokee Swamp in fall foliage.

Bald eagle, *Haliaetus leucocephalus,* in captivity due to a broken wing. This symbol of the United States, once an endangered species, is coming back as nesting in the east has improved.

The Wood stork, *Mycteria americana,* also known as the Wood ibis, is a bellwether of habitat health. In recent years, there has been a precipitant decline in their numbers—shown here in Billy's Lake.

Glossy ibis, *Plegadis falcinellus,* on Merritt Island National Wildlife Refuge.

Fort Raleigh on the Albemarle Sound in North Carolina.

Lake Wawasee, Indiana, was revealed by a retreating glacier following the ice age.

White-faced ibis, *Plegadis chihi*, are normally found west of the Mississippi River.

Mottled duck, *Anas fulvigula*, on the Loxahatchee National Wildlife Refuge, Florida.

Grasses at Bombay Hook National Wildlife Refuge, Delaware.

Fragrant water-lily, *Nymphaea odorata,* in Chesser Prairie, Okefenokee Swamp.

Osprey on the Chesapeake Bay.

Raccoon, *Procyon iotor,* in Silver Springs, Florida.

Big Water in winter — Okefenokee Swamp.

Anhinga, *Anhinga anhinga,* in Loxahatchee National Wildlife Refuge, Florida.

White-reddish egret feeding in the Laguna Madre, Texas.

Dawn at Lake Fausse Pointe, Atchafalaya, Louisiana.

By Way of the Colophon

This limited edition of *Where Water Meets Land* was photographed and written by Lucian Niemeyer. It is privately published by York Graphic Services, Inc., of York, Pennsylvania, for the enjoyment of our clients, friends, and employees. An additional number of this printing is being distributed through the author.

This is the 25th in a series of Keepsake books designed by Ray Chronister and produced under his direction by members of our staff.

The production of this book is completely electronic, with the manuscript supplied by the author on floppy disks. These text files were then ported to our Quark Department where pages were produced using QuarkXPress®. The type face is 10 on 13 Utopia Regular and Laser Plain for display heads. All photographs were 35mm, which were scanned on Scitex 720 Smart Scan and Screen 608 scanner. Low-resolution files were ported to QuarkXPress. High-resolution color proofs were made on the Iris Ink Jet proofer for color evaluation. Color enhancements were made on the Quadra 900 using Adobe Photoshop®. Iris Ink Jet color proofs, with high-resolution images, were used for first page proofs. Final film was output on the Avantra 36 as eight-up imposed film using picture replacement for the low-resolution FPOs. Eight-up Matchprint proofs were made for final approval.

This edition was printed using four-color process, at 175 line, plus PMS 3282 for the fifth color. A spot varnish coating was added to all pictures and the cover was flood coated with aqueous topcoat. The paper is Consolidated Paper, Inc., 100 lb. Centura Dull Plus Text, a recycled paper. The printing was done on our new 40" Heidelberg Speedmaster six-color press, by PrintTech, a division of York Graphic Services, Inc., at 3600 West Market Street, York, Pennsylvania 17404.

All for the joy of doing.